An Easy Proof Of Extraterrestrial Life

And The Discovery Of A Civilization On Mars

Dilip Rajeev

Copyright © 2017 Dilip Rajeev

All rights reserved.

The Image of the surface of Mars on the book cover is sourced from
ESA/DLR/FU Berlin, and used as per https://creativecommons.org/licenses/by-sa/3.0/

The Images indicated as from The ESA in the book are sourced from ESA/DLR/FU
Berlin, and used as per https://creativecommons.org/licenses/by-sa/3.0/

ISBN: 1547234539
ISBN-13: 978-1547234530

Back in 2009, I were on Google Maps exploring the areas around the Sphynx, and the Great Pyramid in the Giza Plateau.

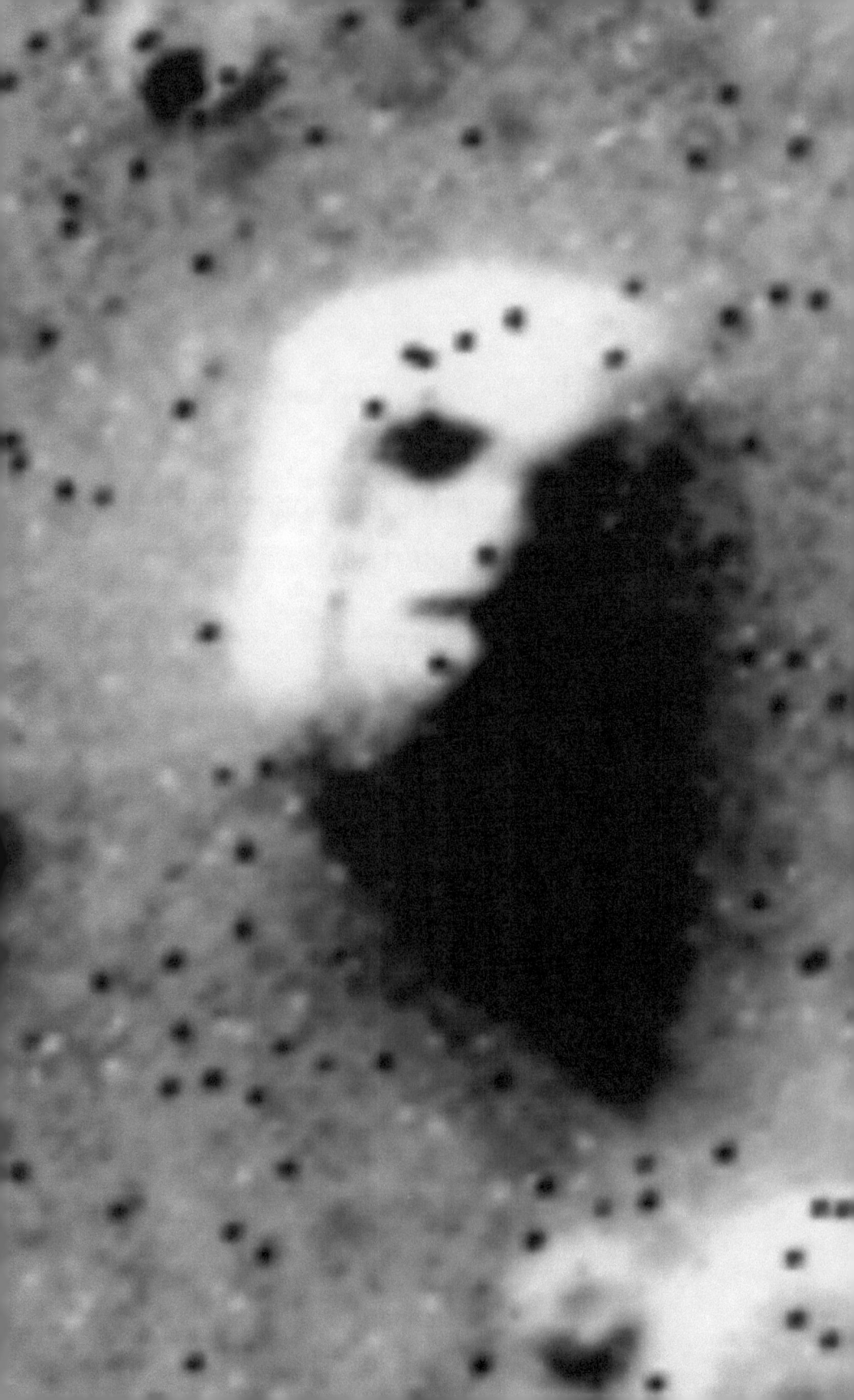

It occurred to me, that a face-like structure - which some scientists had argued is likely artificial - exists on the Cydonia Mensae region on Mars.

ON THE LEFT IS A 1976 Viking 1 photograph of the Face on Mars, http://science.nasa.gov/science-news/science-at-nasa/2001/ast24may_1/

While NASA dismissed the face as an optical illusion, Goddard Spaceflight Center engineers

Vincent DiPietro
and Gregory Molenaar,

argued that the structure resembles a face from any angle

- and that would have the likelyhood of it being artificial, high.

Thinking about this, I turned my attention to the structures near-by.

If any civilization had built something there, it is likely there will be artificial structures of a similar scale near-by.

Artificial structures have geometric characteristics like symmetry, ordering in placement, etc..

For instance, a mud-hill will be rather undefined in form, and its left and right portions will not look similar. In a building, if a draw a line along the center, we often find similarity, a symmetry. We design forms with symmetry, as symmetry is pleasing to vision, and also to serve the engineering design purpose.

An artifact, or artificial design done by a humanoid mind, will thus tend to have symmetry

- say, like a mickey mouse head,

the left and right sides will be similar, and, in the reverse, from such ordering we realize it were designed by intelligent beings.

Buildings, likewise, are placed in an ordering in a city. Roads designed take right-angle turns, rivers don't.

I took a look at the structures near the face thing, for evidence of these.

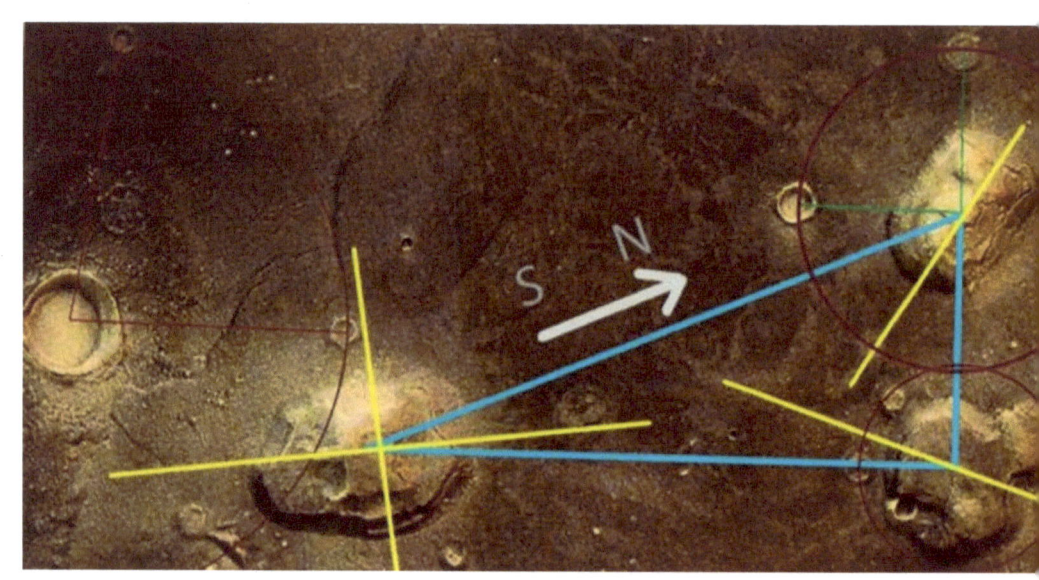

The image to the left summarizes what I observed near the Face On Mars.

The Image of the Area is from the European Space Agency. The sketches on the image were done by the author

IMAGE SOURCE: The ESA , http://www.esa.int/spaceinimages/Images/2006/09/Cydonia_region_colour_image

Structures near the Face On Mars, reveal a high degree of symmetry, and other geometrical characteristics of artificial structures, in their placement.

IMAGE SOURCE: THE ESA, http://www.esa.int/spaceinimages/Images/2006/09/Face_on_Mars_in_Cydonia_region_perspective2

THE SYMMETRY OF THE MICKEY MOUSE HEAD

IS OBVIOUS,

Cydonia Mensae region on Mars. Reveals artificial geometry, typical of engineered artifacts.. Littered with pyramidal structures, and the structures themselves show symmetry, artificial ordering in placement, etc.

IMAGE Source: ESA. Drawings on top, done by the author.

http://www.esa.int/spaceinimages/Images/2006/09/Cydonia_region_colour_image

JUST
AS
THE

AREA AROUND THE FACE
ON
EARTH, THE SPHYNX,

IS LITTERED WITH PYRAMIDS,

THE AREA AROUND THE FACE
ON MARS IS LITTERED WITH
PYRAMIDAL STRUCTURES
BOTH LARGE AND SMALL,

FIVE MAJOR CIRCULAR STRUCTURES IN THE REGION

WE FIND PLACED AT THE CORNERS OF
A

RATHER REGULAR PENTAGON

THERE IS A DEGREE OF SYMMETRY IN THE PLACEMENT OF THE CIRCULAR STUCTURES AROUND THE

BUILDING LIKE STRUCTURES

IF THESE WERE ANCIENT MARTIAN BUILDINGS,

THEY ARE OBVIOUSLY BURIED UNDERNEATH THE GROUND NOW ,

THE FLAT TOPPED PYRAMID

AND THE MICKEY MOUSE HEAD NEAR THE FACE ON MARS,

ARE PLACED ALONG THE NORTH-SOUTH MERIDIAN,

THE THREE STRUCTURES ARE PLACED ROUGHLY ON THE CORNERS OF A RIGHT ANGLED TRIANGLE

We must note that these structures span around a mile across in their lengths, and hence significantly larger a feat of engineering, if artificial, than any structure of the sorts humanly engineered.

I EXPLORED FURTHER

THE AREA AROUND THE SPHYNX AND THE GREAT PYRAMIND ON EARTH

AROUND 5.2 KMS FROM THE SPHYNX,

ON the GIZA PLATEAU,

I DISCOVERED

A BURIED MILITARY ARTIFACT

AT

29 56' 7.16"N 31 6' 29.97"E

IMAGE SOURCE: GOOGLE EARTH

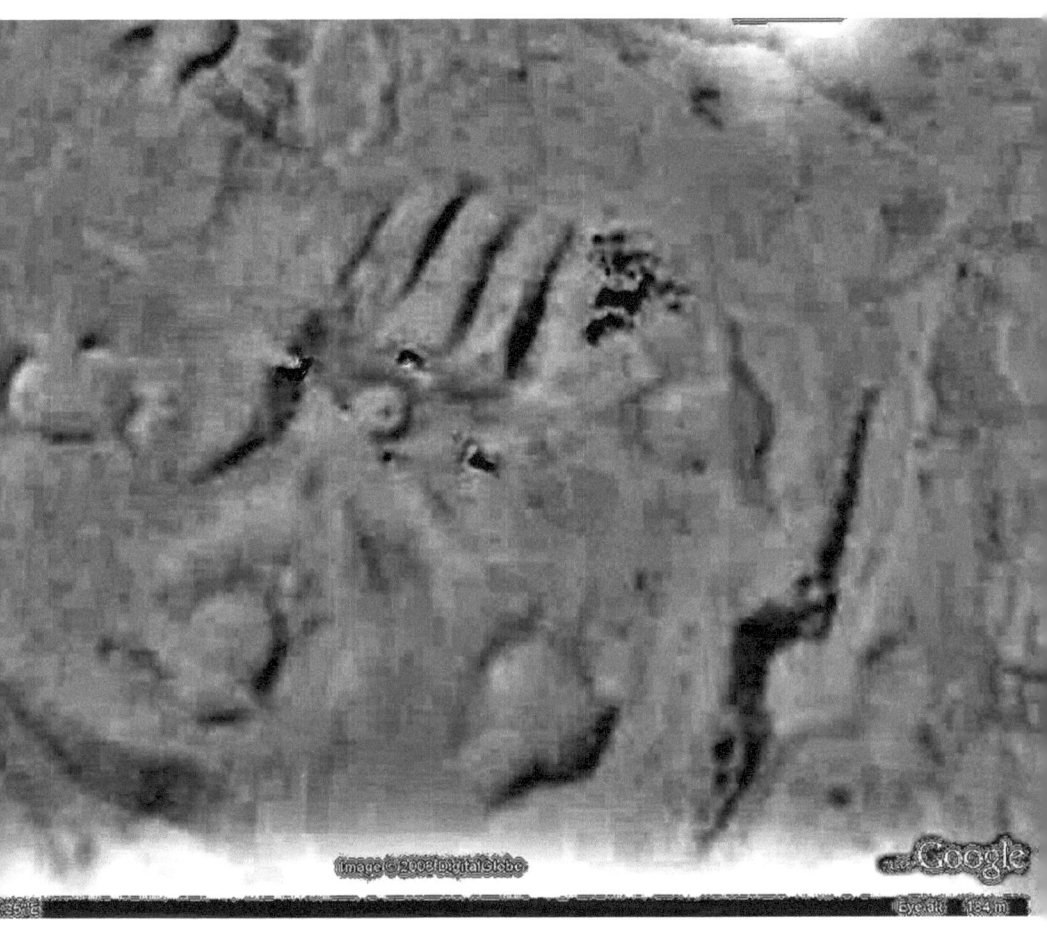

Buried military artifact on Giza Plateau, in the desert region adjacent to the Face On Earth, The Sphynx, Location: 29 56' 7.16"N 31 6' 29.97"E

Dilip Rajeev

IMAGE SOURCE: GOOGLE EARTH

An Easy Proof Of Extraterrestrial Life

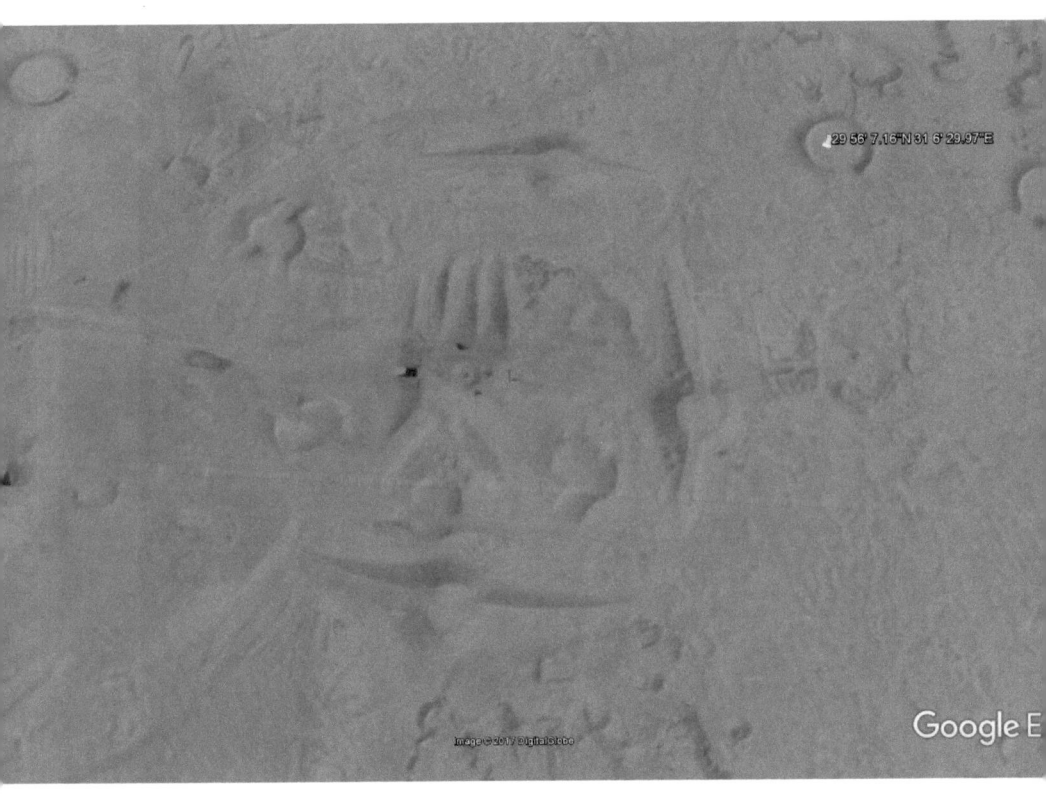

MY INITIAL GUESS WERE THAT THE BURIED STRUCTURE IS DISUSED SURFACE TO AIR MISSILE TECHNOLOGY, NOW UNDER DESERT SANDS

WHAT STRUCK ME WERE THE SIMILARITY IT HAD TO THE FLAT TOPPED PYRAMID NEAR THE FACE ON MARS.

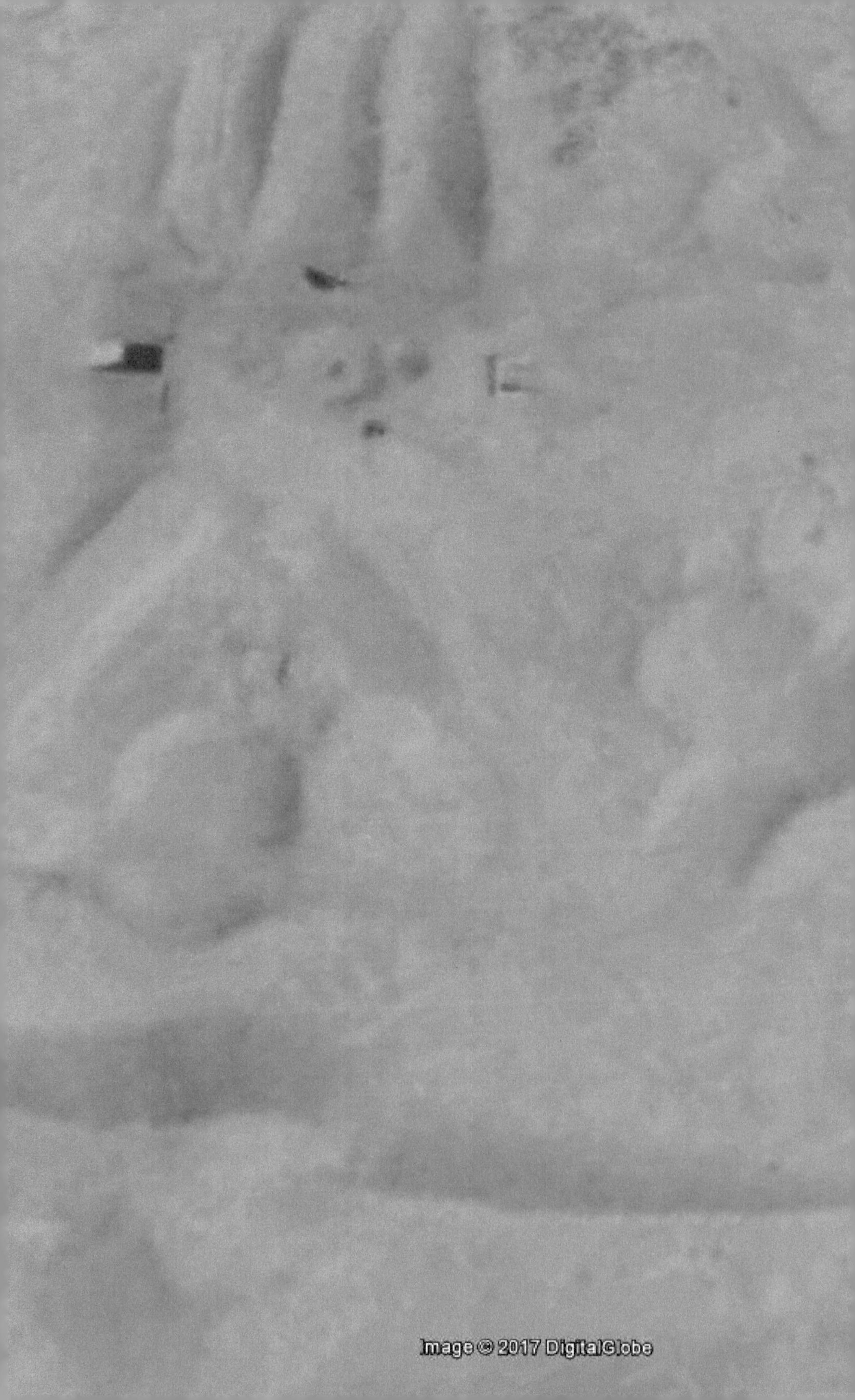
Image © 2017 DigitalGlobe

NOTE THAT THE ARTIFACT ON EARTH HAS RIDGES ON ONE SIDE,

AND A "C" SHAPED STRUCTURE ON ITS TOP,

OPENING AT A 7 O' CLOCK ANGLE

RELATIVE

TO THE RIDGES,

AND NOTE THAT THE MARTIAN ARTIFACT ALSO HAS RIDGES ON A SIDE,

AND A C SHAPED ARTIFACT ON TOP,

OPENING AT A 7 O' CLOCK ANGLE

RELATIVE

TO THE RIDGES,

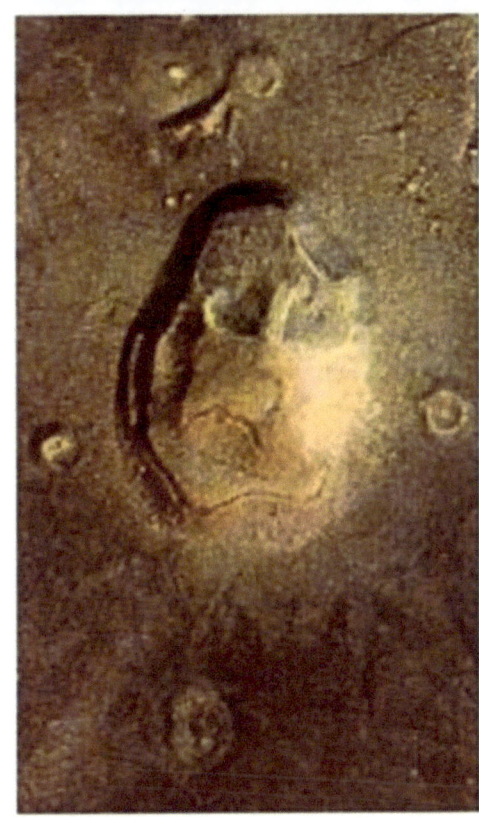

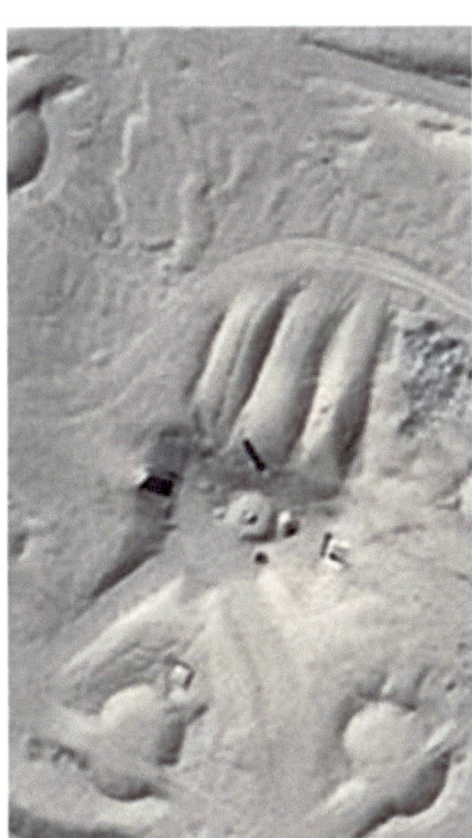

BOTH THESE HAVE A FLAT-TOP PYRAMIDAL DESIGN. AND STAND ON AN ARTIFICIAL PLATFORM.

BOTH THESE HAVE RIDGES ON EXACTLY ONE SIDE

An Easy Proof Of Extraterrestrial Life

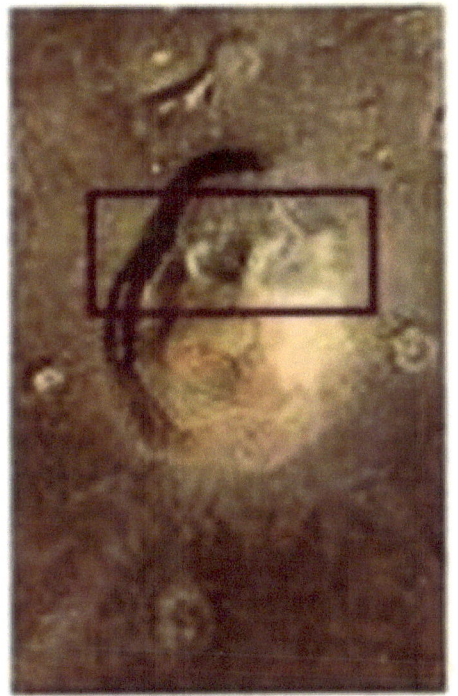

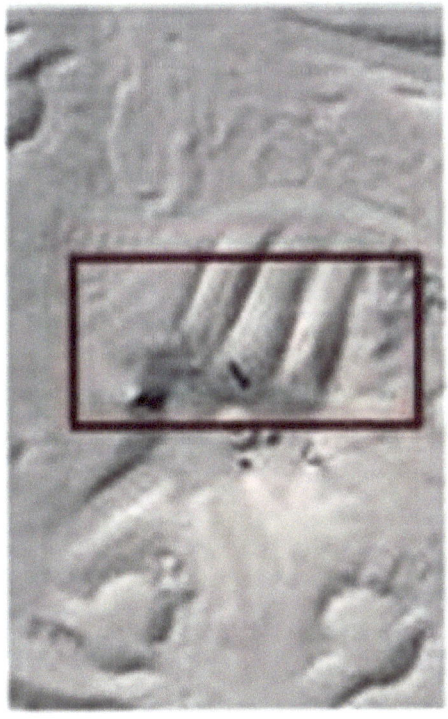

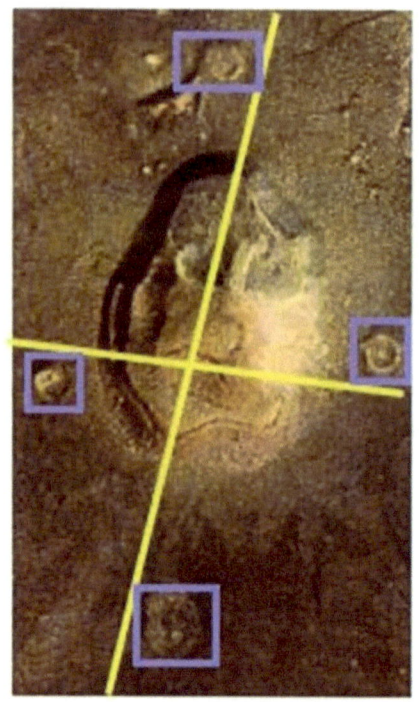

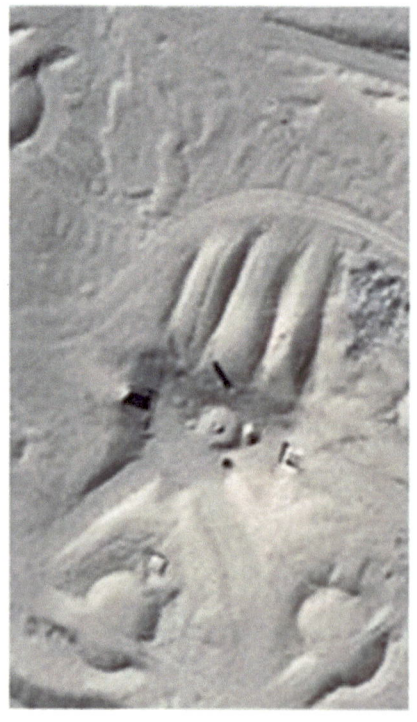

BOTH THE STRUCTURES, THE ONE ON MARS AND THE ONE ON EARTH, HAVE SYMMETRICALLY PLACED CIRCULAR ARTIFACTS AROUND THEM.

Both the structures IN THEMSELVES, show A HIGH DEGREE OF SYMMETRY. Neither is the kind of symmetry in form characteristic of "geographic" forms - of forms carved in mud by random nature - things as rain and wind, nor are rather perfectly and symmetrically placed circular forms around a structure, often the random work of nature.

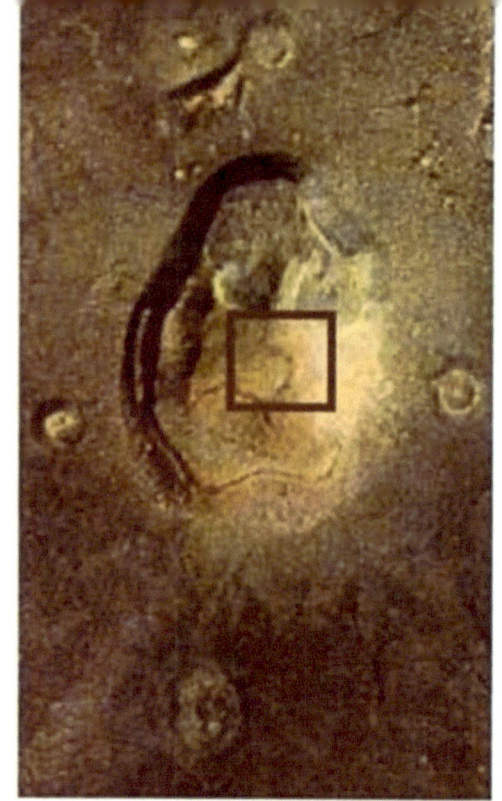

THAT

BOTH

THE ARTIFACTS,

THE ONE ON MARS,

AND THE ONE ON EARTH,

HAVE A C-SHAPED STRUCTURE ON TOP

BOTH OPENING AT

A 7 O' CLOCK ORIENTATION

TO THE RIDGES ON ONE SIDE

IS A LEVEL OF SIMILARITY IN ENGINEERING THAT MAKES IT RIDICULOUSLY OBVIOUS THAT THE TWO ARTIFACTS ARE OF A SIMILAR ARTIFICAL DESIGN

www.ingramcontent.com/pod-product-compliance
Lightning Source LLC
Chambersburg PA
CBHW041112180526
45172CB00001B/223